# Evolution's Irreducible Complexity Problem

Robert P. Waltzer

Seattle      Discovery Institute Press      2024

## Description

Join professor of biology Robert Waltzer as he shows how some evolutionists play a bait-and-switch game. They give examples of microevolution, such as changes in the average beak size of Galapagos finches, and then act as if this proves macroevolution—that is, the evolution of entirely new body plans in the history of life. Not so fast, Waltzer says. An insurmountable obstacle stands in the way of large-scale evolutionary change: irreducible complexity. What's more, your own body is actually an irreducibly complex system of irreducibly complex systems, pointing strongly to intelligent design.

## Copyright Notice

## Library Cataloging Data

*Evolution's Irreducible Complexity Problem* by Robert P. Waltzer

50 pages, 6 x 9 in.

ISBN-13 Paperback: 978-1-63712-058-3, Kindle: 978-1-63712-059-0, EPub: 978-1-63712-060-6

BISAC: SCI008000 SCIENCE / Life Sciences / Biology

BISAC: SCI027000 SCIENCE / Life Sciences / Evolution

BISAC: SCI075000 SCIENCE / Philosophy & Social Aspects

## Publisher Information

Discovery Institute Press, 208 Columbia Street, Seattle, WA 98104

Internet: https://discovery.press/

Published in the United States of America on acid-free paper.

First Edition, First Printing, August 2024.

# CONTENTS

# Evolution's Irreducible Complexity Problem

*Robert P. Waltzer*

Y OU MAY HAVE HEARD IT SAID THAT EVOLUTION IS A FACT. The most reasonable response to such a statement isn't yea or nay. It's, what do you mean by *evolution*? That's because the term *evolution* can mean many different things. It can refer to change over time in the plants and animals that have existed on earth. It can refer to relatively small changes within species. It can refer to the origin of fundamentally new species from earlier forms. It can refer to the common ancestry of all life on earth.

More specifically, it may refer to the theory that natural selection acting on small variations over millions of generations explains the origin and diversity of all life—a theory first propounded by Charles Darwin and Alfred Russel Wallace in the nineteenth century, and further developed and refined in the subsequent 160 years.[1] Finally, the term may refer to some extended version of this theory, one that involves the natural selection/variation mechanism but also other natural mechanisms.

So the term *evolution* can refer to many different things. Being aware of this can help us navigate discussions of evolutionary theory, clear away some of the fog that often bedevils conversations on the topic, and equip us to better assess claims for and against evolutionary theory. To that end, let's briefly unpack a few of the most common meanings of the term: change over time, common descent, natural selection, microevolution, and macroevolution.

## Change Over Time

DIFFERENT PLANT and animal forms existed at different periods of geological history. Nobody seriously questions this claim. For example, if we look at the cat family (Felidae), some species became extinct, such as the Smilodon (known colloquially as the saber-toothed tiger), and others appeared more recently. So members of the cat family were not always the same throughout geological history. This sort of change over time is often described by the term *evolution*. Notice that, by itself, such changes over time say nothing about what drove the change or whether some or all of the different life forms in the history of life share a common ancestry. Change over time is a very modest claim about the history of life on earth, one that few if any question.

## Common Descent

IT'S WIDELY believed that cats—from lions to panthers to domestic cats and everything in between—share a common ancestor. Going beyond the cat family and extending this kind of relatedness to all species, most of Darwin's followers argued that there was a common ancestor for all life on earth,[2] an idea known as universal common descent. This was not a new idea with Darwin, but it gained extra weight after Darwin and Wallace proposed their theory of evolution,[3] and sometimes when somebody refers to "evolution," they mean the idea that all life evolved from a common ancestor. (Note, however, that one can affirm common ancestry without endorsing any particular account of how this occurred. A person might, for instance, think the process was intelligently guided.)

## Natural Selection

IN NATURE, there is often a competition for limited resources. Some members of a species possess this or that helpful variation, which makes them better at obtaining these resources. This, in turn, makes them more likely to survive and reproduce. The ones with the beneficial variations are thus more likely to pass on their beneficial variation to future generations. The beneficial variations are thus selected and the harmful ones filtered out. That's the idea of natural selection in a nutshell.

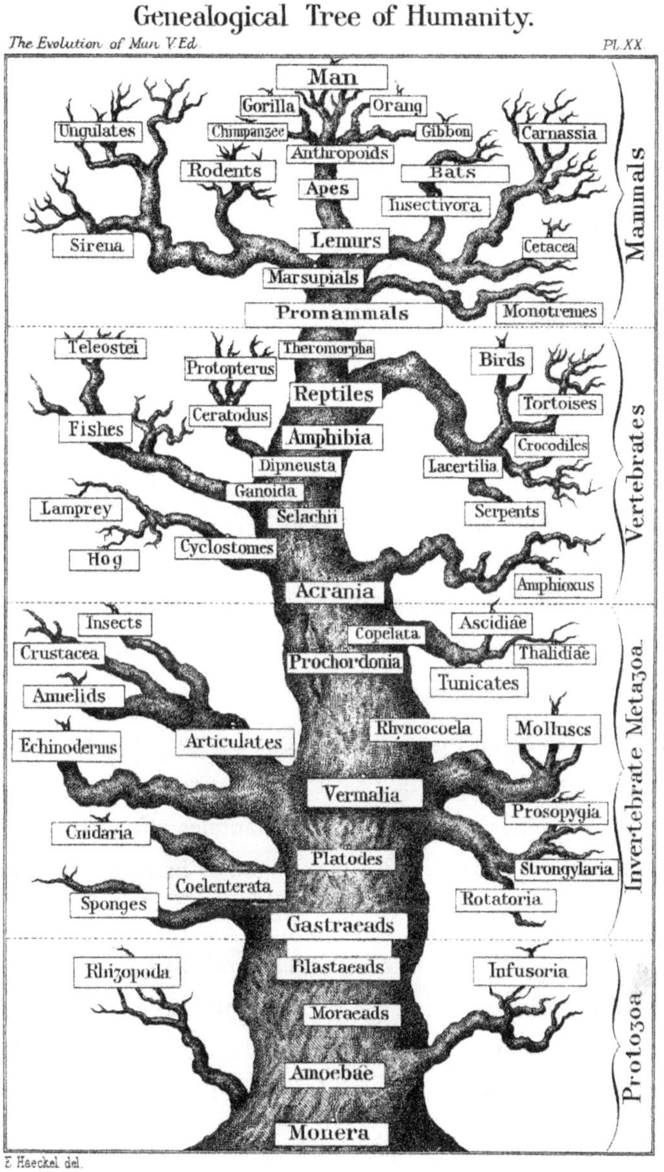

Figure 1. A "tree of life" illustration by nineteenth-century German naturalist
Ernst Haeckel, depicting the idea that all species and groups of species
arose from a common ancestor. Then and now, there is a certain amount of
speculation in the production of such illustrations, and much disagreement
among evolutionary theorists as to what to place where in the tree.

As Darwin noted, when a variation helps a creature survive and reproduce, that variation, called a selective advantage, is thereby more likely to get passed down to offspring and eventually become common in a population.

Let's consider a bird species in an environment where a shift in climate has led to an increase in the number of edible insects living in tree bark, where these bugs are most easily reached with slightly longer beaks than is the average for this species. At the same time, there are fewer seeds to be had in the wake of this climate shift, seeds best broken up and consumed using shorter, sturdier beaks. The beaks will naturally vary in size within this bird population, and those birds with the longest beaks will be better suited for obtaining organisms from the tree bark. So, those birds will be more likely to survive and reproduce in this new climate situation than if they had shorter beaks. The next generation might then have slightly longer beaks on average, and if the climate situation continues, the process will repeat, resulting again in slightly longer beaks, on average, within the population.

On the other hand, birds that live in an environment with seeds hardened by drought might be best adapted with short, stout beaks. This is one of the most commonly cited examples of natural selection acting on random variation, often mentioned in biology textbooks. It nicely illustrates the basic idea of evolution by natural selection.

But notice that the variation is modest, and the beak sizes tend to wax and wane within a fixed range. So, while the example is one of the most commonly used to demonstrate natural selection, it is of only limited use in making a case for evolution by natural selection of fundamentally new animal forms from earlier ones.[4]

## Microevolution and Macroevolution

MICROEVOLUTION REFERS to relatively modest evolutionary changes, like the ones mentioned above. Think of the way finch beaks vary in size, or how some bird species on windy islands without predators adapt to

1. Geospiza magnirostris.
2. Geospiza fortis.
3. Geospiza parvula.
4. Certhidea olivaɔea.

Figure 2. Drawings of Galápagos finches. Short, stout beaks better allow for eating hard seeds. Long beaks better allow for digging in tree bark to catch insects and other small organisms.

become flightless. That's microevolution. Macroevolution, in contrast, means evolutionary change that produces fundamentally new biological structures and forms of life. At some point in the history of life there were no bird wings. And then eventually there were. If evolutionary processes invented bird wings, that's macroevolution.

Entomologist Yuri Philiptschenko is said to have coined the German form of the terms "microevolution" and "macroevolution" in a 1927 German work.[5] The influential Russian-American evolutionary biologist Theodosius Dobzhansky introduced the terminology into English in his 1937 work *Genetics and the Origins of Species*. There he commented that "we are compelled at the present level of knowledge reluctantly to put a sign of equality between the mechanisms of macro- and microevolution."[6]

Why reluctantly? Because as Dobzhansky conceded, while we can directly observe and construct experiments that produce microevolution, we have not been able to observe or produce macroevolution. Even today when we attempt it, more than eighty years after Dobzhansky wrote those words, we continue to hit a microevolutionary wall, and quite early. We see this even in the case of microbes that reproduce rapidly and in enormous populations, where millions of mutations and tens of thousands of generations are possible.

So what "compelled" Dobzhansky to equate micro- and macroevolution? It seems that it was his commitment to the existence of macroevolution combined with the confessed lack of direct observational support for macroevolution. That is, he wanted to demonstrate that the entire diversity of life required no mechanisms beyond those of microevolution, but he couldn't observe or produce macroevolution in order to clinch the case, so he settled for demonstrating microevolution and then equating it with macroevolution.

Dobzhansky and other evolutionary theorists are, of course, free to equate the two, but the rest of us are free to note the crucial difference, the very inequality that Dobzhansky himself noted. Indeed, reason urges us to do so. Acknowledging the crucial difference, the inequality, is key to clear thinking on the issue.

Can unguided microevolutionary changes accumulate and lead to dramatic macroevolutionary changes? Can they, for instance, lead to the rise of the first ocean-going animals, or the first dinosaurs, or the first mammals? Can they produce sonar in bats? We cannot even rationally address the question until we first acknowledge the clear difference between microevolution and macroevolution.

## Equivocation Alert

EQUIVOCATION IS the logical fallacy of changing the meaning of a word in the middle of an argument. Take, for example, the following argument: "Whenever it's sunny out, the streets are dry. Harriet has a sunny disposition. Therefore, whenever Harriet is out, the streets are dry." The word "sunny" in the first place means the sun is shining; in the second place it means upbeat and cheerful.

Arguments defending evolution often resort to equivocation, trading on the fact that the term *evolution* can refer to several different things. Take, for instance, someone who says that evolution is a fact, where by "evolution" he means the idea that natural selection working on accidental genetic mutations has produced macroevolution, the evolution of new structures and forms in the history of life. Suppose that this person then tries to prove evolution in this sense by citing how breeders have bred different kinds of dogs—microevolution by artificial selection. So, at first evolution means macroevolution by natural selection, but then evolution refers to microevolution by artificial selection. The evolutionist has used a term in his argument in two different ways, but acts as if the meaning hasn't changed, or at least hasn't changed significantly. This is an example of the fallacy of equivocation.

We must understand that those who equivocate with the term "evolution" may not be acting in bad faith. They may not realize they are equivocating. But that doesn't mean one has to go along with their confusion. Instead, the next time someone asserts something about "evolution," press "pause" and ask what exactly the person means. And if the person points to evidence of evolution in action, ask yourself, what kind of evolution, if any, was just demonstrated? And does that evidence clinch the case for full-blown, unguided, microbe-to-man evolution? Does it contribute to the case but not clinch it? Is the provided evidence entirely beside the point? If so, what else, if anything, needs to be demonstrated to clinch the case for the full-blown theory of evolution?

## A Competing Explanation

IN ASSESSING the claims for modern evolutionary theory, a worthy goal is to proceed as reasonably as possible, and to follow the evidence wherever it leads. Equivocating with the term evolution doesn't get us there. It produces fog and confusion when what we want is clarity and insight.

How can we proceed reasonably and in a way that's guided by the evidence? The science of biological origins is a historical science, and the historical sciences have developed a shared methodology for doing just this, proceeding reasonably and in a way that's guided by the evidence.

Historical sciences study clues in the present to solve mysteries about the past. In the historical sciences, investigators compare competing explanations for a given event or set of clues about a past event. Sometimes two or more explanations seem to adequately explain an event in question. If a decisive tie-breaking clue can be found, then one explanation can replace the other. If not, then it is not right for one explanation to declare itself "the truth" and anything else false.

Take an example from the historical science known as forensics, used to study crime scenes. A man is found dead in his home. Forensic scientists are called in to investigate and determine the cause of death. Death by some natural illness? By murder? By suicide? By accident? The man was found in his tub with evidence of a hard blow to the back of his head. One of the investigators concludes the man slipped and struck his head on the side of the tub. You can't very well hit the back of your own head hard enough to kill yourself, the investigator reasons. So clearly it was not suicide, he announces. Instead, it was death by accident.

But a second investigator points out that the man might have been struck in the back of the head by someone else. Maybe it wasn't an accident. The first investigator points to the blow to the back of the head and notes that the blow is perfectly consistent with his theory of an accidental fall. Case closed, he says. When the other investigator raises an eyebrow, the first investigator accuses the other of sensationalism, of preferring the most dramatic explanation, murder. There is no evidence of forced entry into the house, the first investigator further notes, and if every bathroom accident is to be attributed to some mystery murderer, where will the conspiracy theories end? No, he insists. He's going to write this up as a simple accident. The homeowner slipped, hit his head, and died. End of story.

Is that any way for a forensics expert to proceed? Of course not. He's failing to give due consideration to the murder hypothesis. Instead, he should acknowledge that there is an alternative explanation and that this is not an open-and-shut case. Further evidence is required to establish one view over the other.

The first investigator mentioned that there was no sign of a forced break-in. Good, but this is hardly decisive. The man could have been murdered by a friend or family member. Or maybe he left the front door unlocked. Instead of rushing to judgment, the first investigator should scour the scene more carefully for additional tie-breaking clues that might prove genuinely decisive one way or the other. If he did, he might find, for instance, a pewter bookend in the hallway that shows traces of having recently been cleaned with bleach, while the second bookend has not. The bookends also are shaped in a way that matches the wound on the back of the dead man's head. Important clues, but they won't even be considered if the first investigator has his way.

There is a case in origins biology not unlike the fictional scenario above. Similarities in genetics can be observed in various species. For instance, there is a set of genes involved in specifying the organization of the body as it develops. These are called Hox genes. We can see similarities in some Hox genes in species as disparate as fruit flies, octopuses, and humans. Evolutionary theorists assert that common ancestry is the explanation for these genetic similarities. That is, the similar genes are said to have passed from a common ancestor to these descendants, the various forms that share these common genes. Okay, that is one possible explanation for genetic similarities, but it isn't the only explanation.

Just as software developers reuse lines of computer code in different contexts as they design new software programs, or car designers reuse the principle of four wheels and two axles when designing a new car, so too might a designer of life have reused lines of genetic code across different species. This too is a potential explanation.

Indeed, computer scientist Winston Ewert argues that the pattern of similarities and differences in various genomes has more in common with the pattern of similarities and differences we find among the work of software designers, who reuse software modules in different contexts even as they build software programs with new elements and new arrangements of existing modules. Ewert suggests that this pattern of similarities and differences favors purposive design as the primary cause of the genetic programs that help code for the diversity of biological forms we find in the biosphere.[7]

Some evolutionary theorists, unfortunately, are like the first investigator above who dismissed the murder hypothesis by ruling it out as overly sensational, and then simply reiterated that his death-by-accidental-fall explanation was consistent with the blow on the back of the victim's head. These evolutionary theorists insist that inferring purposive design "isn't science" and behave as if common descent via unguided evolution is *the* explanation for genetic similarities, end of discussion.

## Junk DNA or Junk Argument?

FORTUNATELY, SOME evolutionary theorists do a bit better than this. They argue that Darwinism's process of blind evolution by trial and error could be expected to produce a lot of genetic junk, and that our genomes are riddled with junk DNA. This, they say, is expected on evolutionary grounds but not if life were the work of an intelligence.

Francis Collins and Karl Giberson offer the example of a gene involved in synthesizing vitamin C. "Primates, including humans, require vitamin C in their diet, or they will suffer a disease called scurvy," they write. "What happened? The human genome has a degenerated copy of the gene that makes the enzyme for synthesizing vitamin C. This 'broken' gene has lost more than half of its coding sequence. To claim that the human genome was created by God independently, rather than having descended from a common ancestor, means God inserted a broken piece of DNA into our genomes. This is not remotely plausible."[8] Such a broken gene is called a "pseudogene."

There are, however, answers to this argument. Genetic errors could easily be from devolution over time. That is, an original design without these errors took on mutational errors in the course of numerous generations. While the replication process is extraordinarily accurate and even contains an astonishingly sophisticated error-correction system, it isn't wholly error-free, so we might expect errors to accumulate in some parts of the genome.

Also, because chimps and humans are similar in structure and function (they are both primates, after all), it is not unexpected that they would have a similar genetic sequence in a similar location. The fact that this gene ceased to work within the human lineage is not evidence for a common ancestor with chimps, let alone evidence for common ancestry instead of common design.

Also, the fact that our supposed pseudogene does not function for vitamin C production does not mean that it does not have any function. The version in humans could still have some as yet undiscovered function. The Darwinian paradigm has discouraged the search for further function in this and many other genes whose possible function has yet to be determined.

There is also the question why the remaining non-mutated portion is intact. If the gene were totally useless, why wouldn't the remainder of it eventually undergo further mutation or even be deleted from the genome? I suspect it does not undergo such mutation or deletion because that might be deleterious to the individual, decreasing her chance of survival—because the gene may serve a function. And therefore the unmutated portion remains, serving some still undetermined but necessary function.

Finally, if the evolutionary story that Collins, Giberson, and others tell were the case, we could expect to see this confirmed by the wider pattern of phylogenetic (evolutionary) relationships among the various animals found to lack the ability to synthesize vitamin C. But just the opposite is the case. As Sebastian J. Padayatty and Mark Levine, writing in the journal *Oral Diseases*, note, many of "those animals that lack vitamin C synthetic ability do not bear any phylogenetic relationship to each other, implying many independent mutations all resulting in the same phenotype. No common environmental influence is apparent. To date, there is no satisfactory evolutionary explanation for the apparent random loss of vitamin C synthetic ability."[9]

As for the vast sections of the genome that evolutionary theorists have deemed "junk DNA," there are developments that would seem to favor the design hypothesis over that of blind evolution. Design theorists predicted years ago that much of what evolution proponents designated as junk would turn out to have important functions,[10] and this design prediction has already proven true. Moreover, every year scientists uncover more evidence of the previously unknown functions of this so-called junk.

Ann Gauger, Ola Hössjer, and Colin Reeves comment that "pseudogenes have not received much attention in the scientific literature because they are assumed to be 'junk.'" But they say this is changing. "Where pseudogenes have been carefully studied, they are often found to be functional, and in some nonstandard ways," they write. "Part of the problem is that a pseudogene may be active in specific tissues only during particular stages of development, making identification of their functions difficult. Nonetheless, researchers in the field are confident that continued research will yield more evidence of functionality."[11]

## Darwin's Test

BESIDES THE issue of genetic similarities and differences, there are other potentially tie-breaking clues worth considering, evidence that might point toward the work of a creative intellect and away from blind evolution. Let's look at one of these next, see how Darwinists respond, and consider how one might respond in turn.

In his *Origin of Species* Darwin offered a way to test and possibly even falsify his theory of evolution. "If it could be demonstrated that any complex organ existed which could not possibly have been formed by numerous, successive, slight modifications," he wrote, "my theory would absolutely break down."[12] With the development of high-powered microscopes and new observational techniques, we now know of many biological structures at the molecular level that can serve as candidates for potentially falsifying Darwin's theory. In the past few decades, numerous tiny biological structures have been discovered, intricate structures often referred to as *molecular machines*. Michael Behe, professor of biochemistry at Lehigh University, has argued that at least some of them could not have evolved in the way Darwin envisioned and therefore do falsify Darwinism. He suggests that a better explanation for their origin is purposive design; that is, a designing intellect fashioned them.

Central to Behe's argument is the idea of irreducible complexity. He defines irreducible complexity as "a single system composed of several well-matched, interacting parts that contribute to the basic function, wherein the removal of any one of the parts causes the system to effectively cease functioning."[13] If such molecular machines exist, how could they have evolved one small step at a time, given that they don't work until all of their many essential parts are in place? Behe argues that they couldn't have, and that molecular biology has turned up several irreducibly complex molecular machines, marvels of nano-technology that could not have evolved in the mindless, gradual way that Darwin and his successors envisioned.

To grasp the concept of irreducibly complexity, it helps to envision a familiar machine, the common mousetrap. It's relatively simple as far as machines go, but as Behe notes, it still requires "several well-matched interacting parts"[14] in order to function properly. If the hammer is removed, then the mouse will not be trapped. If the holding bar isn't there to hold the hammer back, then the trap will be closed all the time and won't catch anything. Take one of these essential parts away, reduce it by one key part, and it's no longer a functioning mousetrap. This is what Behe means by *irreducibly complex*.

**IRREDUCIBLE COMPLEXITY**

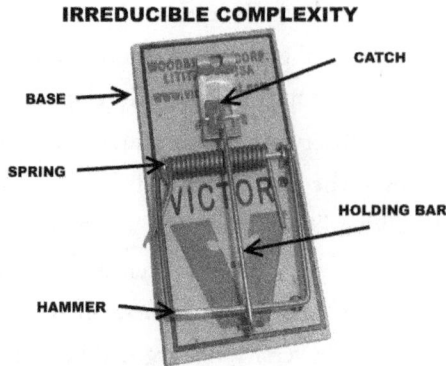

Figure 3. A mousetrap as an example of irreducible complexity.

## The Little Engine That Could

Now LET's move from the relative simplicity of the mousetrap to examples from biology. Behe gives several in his book *Darwin's Black Box.*[15] These include the human blood-clotting system and light-sensing mechanisms in our eyes. But let's focus on his most famous example, a miniature motor that in some ways puts to shame even the motor of a winning Formula 1 racecar.

In an electron microscope the bacterial flagellum is a long whip-like tail extending off the jelly bean-shaped bacterium (Figure 4.4). But it's so much more. We now know that this whip-like flagellum functions as a motor, spinning many thousands of times per minute and moving the bacterium as a propeller moves a boat through the water.

Similar to a man-made machine, the motor component of the flagellum has a drive shaft attached to a rotor that turns within a stator and is anchored by bushings. Unlike a boat motor, however, it's fueled by hydrogen ions rather than gasoline.

Using advanced electron microscopes, biologists have been able not only to see the flagellum but also to learn about the tiny parts of this remarkable nano-machine. We can see some of them in the sketch below (Figure 4.5). Keep in mind that the actual flagellum is even more sophisticated than this simplified drawing suggests.

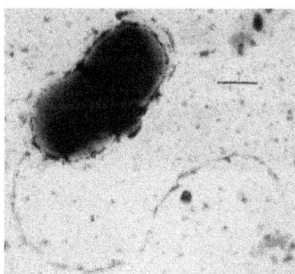

Figure 4. Electron micrograph of a bacterium. The large oval structure is the bacterium, and the long whip-like tail is the flagellum.

What does the bacterial flagellum have to do with evolutionary theory? For modern Darwinism to be true, this remarkable nano-machine had to have evolved through "numerous, successive, slight modifications," with no intelligence involved, only a series of small random mutations sifted by natural selection, over thousands or millions of generations.

Each mutational step has to be small because the odds are too great to produce big mutational improvements in a single big leap. (This is the standard view of neo-Darwinists.) In the same way that you could never randomly dump a box of Scrabble letters onto the board and expect

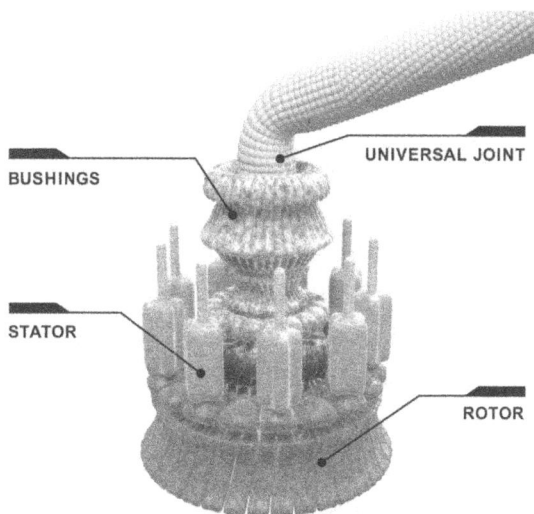

Figure 5. Depiction of the key components of the bacterial flagellar motor.

them to neatly spell out a completed game of interlocking English words, we cannot expect random genetic mutations to suddenly arrange DNA into the proper coding for the entire bacterial flagellum. No, taming the odds requires advancing in baby steps.

In addition, each of these baby steps on the evolutionary pathway must be functional to provide a selective advantage and get passed on to the next generation. It's not enough that the mutation will one day prove useful. Evolution doesn't look ahead. Evolution is blind. That's why the evolutionist Richard Dawkins called it "The Blind Watchmaker" in a book by that title. The process doesn't look ahead and think, "Hmmm, wouldn't it be handy for a bacterium to have a tail that can propel it. I think that over the course of many generations I'll locate and assemble a few dozen distinct, very specialized parts into a snazzy motor." No, evolution doesn't look ahead. It doesn't think. There's just a small random mutation in the DNA. Then the mutation either helps the organism or it doesn't. And it either gets passed on to another generation or it doesn't. Whether the mutation will one day be useful when assembled with various other possible mutations is of no concern to the evolutionary process. It doesn't care about the future. All that matters to it is whether a new mutation helps with function then and there, with survival and reproduction. It's blind to everything else.

Because of this, evolution requires a series of tiny, functional steps from simpler ancestor to fully outfitted flagellar motor. But there doesn't appear to be such a pathway to the bacterial flagellum. Instead, the evidence suggests that the entire structure must be present, with all the parts in place, for it to propel the bacterium. In this way it's just like the mousetrap. If it only has some of its several essential parts, it doesn't work as a flagellum. If one or more of these parts is a dud, or missing, the entire flagellum grinds to a halt. Or if all the parts are present and in good working order but haven't been precisely arranged, again, the flagellum doesn't work and instead just soaks up valuable resources from the bacterium, decreasing its likelihood for survival.

But of course the bacterial flagellum *does* work. How did such a technological marvel arise? Supporters of modern evolutionary theory have tried to provide a story as to how a series of lucky coincidences could evolve such a system. One idea they proffer is what's called co-option. That is, nature co-opts earlier simpler molecular machines on the way to creating the bacterial flagellum. This is a creative idea, to be sure, but even just in principle it faces a serious shortcoming. The parts on the simpler machines would have to be reworked to function in the bacterial flagellum and somehow all be fitted into place—just as a garage tinkerer would have to rework and carefully assemble scrap parts from old lawnmowers and such to build a motorized go-kart.

Evolutionists have proposed one possible precursor machine for the bacterial flagellum, a needle complex known as the type III secretory system (TTSS). Did the TTSS help pave the evolutionary pathway to the bacterial flagellum? The notion has several problems, as Scott Minnich and Stephen Meyer explain:

> This argument seems only superficially plausible in light of some of the findings presented in this paper. First, if anything, TTSSs generate more complications than solutions to this question. As shown here, possessing multiple TTSSs causes interference. If not segregated one or both systems are lost. Additionally, the other thirty proteins in the flagellar motor (that are not present in the TTSS) are unique to the motor and are not found in any other living system. From whence, then, were these protein parts co-opted? Also, even if all the protein parts were somehow available to make a flagellar motor during the evolution of life, the parts would need to be assembled in the correct temporal sequence similar to the way an automobile is assembled in factory. Yet, to choreograph the assembly of the parts of the flagellar motor, present-day bacteria need an elaborate system of genetic instructions as well as many other protein machines to time the expression of those assembly instructions. Arguably, this system is itself irreducibly complex. In any case, the co-option argument tacitly presupposes the need for the very thing it seeks to explain—a functionally interdependent system of proteins. Finally, phylogenetic analyses of the gene sequences suggest that

flagellar motor proteins arose first and those of the pump came later. In other words, if anything, the pump evolved from the motor, not the motor from the pump.[16]

Where does this leave us as regards the bacterial flagellum motor? So far, no one has proposed a reasonable step-by-step plan for how such a thing could have evolved. No one has even come close. So, as Behe notes, Darwin's theory has been put to Darwin's own test and failed.

The bacterial flagellum isn't alone, either. There are numerous complex systems for which no one has come even close to showing how they could have evolved through a series of slight, successive modifications. Indeed, evolutionists have not succeeded in providing a detailed, workable evolutionary pathway for any irreducibly complex biological system. Not one.

Hypothetical stories starved of evidence or detail are not adequate. Without a reasonable explanation for how such systems could have gradually evolved through purely natural mechanisms, modern evolutionary theory fails to explain a very significant part of what we see in biology and, indeed, fails one of its most basic tests.

## Irreducible Complexity in Our Own Bodies

LET'S MOVE from tiny biological machines to large-scale systems in the human body that display irreducible complexity—in particular, the systems for transporting blood and carrying oxygen to the tissues. Such work requires the coordination of multiple structures and subsystems, all with extremely fine-tuned specifications. We need a heart and a series of blood vessels, making up the cardiovascular system. And we need lungs, airways, and muscles to bring air into the body—the respiratory system. The cardiovascular system's many essential parts must be in the correct location and arrangement for the system to work, and to get oxygen to the tissues it must perform an exquisitely complex dance with the respiratory system, which also needs its many essential parts in place.

There actually are many other subsystems that must be in place and working with these, but here we will focus on these two systems. They are crucial because the cells of your body need oxygen to keep them alive and functioning.[17]

The respiratory system contains lungs and airways as well as important muscles, including the diaphragm. When the diaphragm contracts, air is pulled into the lungs through the airways. The lungs, the airways, and the muscles must all work together for the respiratory system to function. But once the air enters the lungs, nothing would happen if oxygen couldn't get into the bloodstream. So there needs to be a precisely structured interface between the respiratory system and the cardiovascular system. Each system needs the other.

The tiny grape-like sacs in the lungs that receive air are called alveoli or alveolar sacs. The tiny blood vessels surrounding these sacs are called pulmonary capillaries. When blood first enters the lungs, it is low in oxygen and bluish in color. After absorbing oxygen, the blood turns red.

Together the alveolar sacs and capillaries form the interface by which the oxygen gets from the airways into the bloodstream. The walls of the alveoli and the walls of the capillaries have to be close together and incredibly thin for oxygen to diffuse efficiently. And there have to be a sufficient number of alveoli and capillaries for enough oxygen to diffuse. All these must be very precisely structured to work.

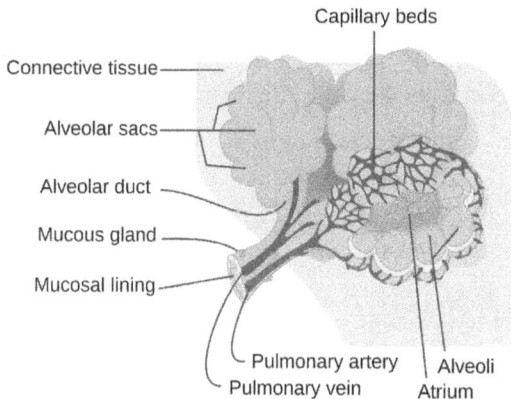

Figure 6. A highly simplified drawing of alveolar sacs and pulmonary capillaries.

## Oxygen's Beast of Burden

YET EVEN with a complete cardiovascular system, a complete respiratory system, and a complete interface between them, there would still not be enough oxygen carried to the tissues. Why not? Because oxygen has very low solubility in the blood fluid (plasma), so the blood would not be able to absorb and transport sufficient oxygen to keep us alive. To get enough oxygen to the cells we need almost fifty times the amount of oxygen that naturally dissolves in the plasma.

This is a challenging biological engineering problem. What are the options for solving this problem? One option would be to enlarge each of these systems to absorb and carry more oxygen. Unfortunately, this would create a new problem because each part would have to be many times bigger. The heart would be bigger than the chest. And the lungs would be even larger. And where would all the blood vessels and the airways be located? There would not be enough room. Enlarged systems would not work. There needs to be a way to get fifty times more oxygen to dissolve in the blood without substantially enlarging the heart, lungs, and blood vessels. Even the smartest chemist might be stumped by this problem.

How was the problem solved? The answer is in a molecule called hemoglobin.

Hemoglobin is a protein and, as are all proteins, it is coded for with a twenty-character alphabet of amino acids. The comparison to language is illuminating but also in need of qualification. Amino acids may not function by symbolizing something else, in the way human words do, but are more akin to precisely specified parts in a sophisticated machine component. However, they are identical to software or text in English, in the sense that the choice and order of the "letters" are crucial.

Hemoglobin has 574 amino acids contained within four chains. As the letters in this paragraph are precisely selected and ordered to communicate information about hemoglobin, so the various amino acids are ordered within hemoglobin's four chains to give the molecule exactly the right shape and function. Hemoglobin, similar to thousands of other proteins, is an extremely complex and amazingly constructed entity, fine-tuned to carry out a crucial function. One molecule of hemoglobin can bind four oxygen molecules, hold onto them, and release them where and when they are needed. As many as 280 million hemoglobin molecules are found within one red blood cell.[18] There are about five trillion red blood cells per liter of blood, and approximately five liters of blood in the average adult human. By binding four oxygen molecules to each hemoglobin molecule, and by having so many hemoglobin molecules in a red blood cell, and so many cells in our blood, we are able to deliver enough oxygen to our tissues to stay alive.[19]

And proteins, keep in mind, often have very limited tolerance for error in their sequence of amino acids, with random mutations usually leading to diminished or lost function. Indeed, there is growing laboratory evidence that the specific and highly improbable sequence of amino acids that code for a particular fold (functional shape) of a protein can only tolerate a few alterations before losing function. For instance, one study reported on in *The Journal of Molecular Biology* made the case that if left to chance, the odds of a protein 153 amino acids in length having just the right sequence to fold and perform a specific function is about 1 in $10^{77}$.[20] That's one chance in 1 followed by 77 zeroes. (More on those findings below.)

## Putting It All Together

THE CARDIOVASCULAR system is an exquisitely orchestrated marvel, but unlike a marvelous orchestra, it is not merely diminished by the absence of one of its instruments—one of its subsystems. Instead, it would cease to work altogether. We need all of them present, all of them working, and all of them carefully coordinated to get oxygen in sufficient amounts to the numerous parts of our body.

No one has ever been able to offer a detailed account of how the cardiovascular system could have gradually evolved through blind natural forces. But even if a fully functional cardiovascular system could somehow have evolved in this way, it would still not be enough. Remember Darwin's challenge: "If it could be demonstrated that any complex organ existed which could not possibly have been formed by numerous, successive, slight, modifications, my theory would absolutely break down." What applies to a single organ applies equally to a system of interdependent organs, ones that cannot survive and function without the others. The system for providing oxygen to our cells requires the cardiovascular system with all its many essential parts, the respiratory system with all its many essential parts, the precisely structured interface between them, and the hemoglobin molecule with its special binding properties to be able to carry oxygen molecules in sufficient numbers.

None of these systems and subsystems are optional. Without all of them in place at the same time, we wouldn't be alive. So nature cannot evolve a few of the parts or subsystems at a time and wait around for countless generations while some of the other parts and subsystems evolve and come on line. Nature can't even wait around a single generation. It's do or die. Evolution, in this case, would need to proceed not by tiny steps but in a gargantuan creative leap, like dumping boxes and boxes of Scrabble letters on a table and somehow they spell out a series of interlocking English words that, in turn, spell out the unfolding sentences of an exquisitely ordered script for a multi-act play. There are less complex cardiovascular/respiratory systems in the biosphere, of course, but these have their own internal logic, and no one has been able to offer a detailed, functional evolutionary pathway moving from these relatively simpler systems (though still almost unimaginably sophisticated by human engineering standards) to a cardiovascular/respiratory system such as we possess.

## Neutral Evolution

In the face of the irreducible complexity challenge, evolutionists have proposed other scenarios (in addition to the idea of co-option discussed above) to explain how such complex systems might have arisen by undirected, accidental processes. For example, some have argued that *neutral evolution* might help explain the origin of new biological functions. In simple terms, neutral evolution is based on the observation that some mutations in DNA are neither harmful nor helpful, but are neutral and don't seem to impact the organism in a significant way. As a result, natural selection would neither weed out nor select for these neutral mutations and, proponents claim, the mutations could continue to accumulate until at some later point they happen to join together in a lucky way to positively help the organism.

One of the most popular suggestions is that perhaps a gene could have been duplicated and one of the copies was "experimented on" through random neutral mutations while the other copy maintained the original function required for the organism to survive. Presumably the new copy would eventually acquire the more complex functional role and the old copy could then be eliminated.[21] It's certainly an interesting idea, but some problems with it have been identified:

1.  Any step-by-step scenario that has been proposed is completely imaginary. The evidence is lacking to support the step-by-step evolution of the new system.

2.  Neutral evolution avoids the problem of a given DNA sequence having to be functional at every step, but at a steep price. Now the process must build without natural selection to preserve beneficial changes. There is nothing in neutral evolution to lock in mutations that are building toward the completed sequence required by the system or subsystem in question. Remember, neutral evolution is said to happen where fitness considerations are mostly ignored, particularly as regards potentially helpful mutations. As a result, neutral evolution abandons natural selection in favor of random outcomes, i.e., luck. Yet chance is a poor candidate for building the kinds of irreducibly complex systems found throughout biology. As the required number of units in a sequence grows, the odds against it happening by chance quickly grow astronomical, becoming

essentially impossible long before the sequence is long enough to code for a functional system. So although neutral mutations might accumulate in an organism's DNA over time, those neutral mutations do not explain how new and sophisticated biological features arose.

3.  There is a high energy cost for an organism to experiment with something that does not immediately produce a benefit. Evidence exists showing that some organisms will be quick to inactivate genes that have no adaptive/selective advantage.[22]

## Like a Weasel

A PROPONENT of evolutionary theory might concede that this or that evolutionary scenario may have its shortcomings, but argue that at least we know that random point mutations in DNA can do impressive things at the level of individual genes and proteins. If that's the case, then surely evolution can gradually accumulate new lines of functional genetic code and, with those, new functions, features, organs, species, and ultimately, even whole new body plans. Let's consider the first half of this claim, since if modern evolutionary theory should work anywhere, it should be at the modest level of genes and individual proteins.

In the dance of molecular biological life, there are many different kinds of genes and many different kinds of proteins. In simplified terms, leaving RNA aside for now, let's assume that a gene is a stretch of DNA that can code for a given kind of protein.[23] The information content in DNA is based on the arrangement of the four letters (bases) in DNA. The letters are held in place by other molecules and organized into a special spiral called a double helix. The arrangement of the bases in DNA is comparable to the arrangement of letters in the sentences and paragraphs of this book, except that in the case of the English language there are twenty-six letters, while in DNA there are four letters (bases), typically abbreviated as A, T, G, and C. In both cases, it is the specific order of these letters that provides meaning or function. It isn't enough just to have all these DNA letters piled in. To function they have to be in a particular, functional sequence, in the same way that the letters of the sentences in an instruction manual have to be in some particular order to have meaning.

What is crucial about the information found in DNA? It is involved in specifying the arrangement or sequence of amino acids found in proteins, which were discussed a little earlier. Again, there are twenty different amino acids used in the synthesis of proteins. Connecting what was already said about proteins with DNA, the four-character alphabet found in DNA codes for the twenty-character amino acid alphabet. For instance, the hemoglobin molecule discussed above, which has 574 amino acids, has to have the right amino acid letters in the right order so that it will fold and function properly. The correct amino acid placed in the proper sequence depends, in part, on genes, and the information in the genes is based upon the ordering of the four DNA letters—A, T, G, and C.

In the previous century scientists deciphered the genetic code so that we now know how the cell takes the information in DNA (the sequence of the bases) and transforms it into a sequence of amino acids in proteins. The process is quite complicated.[24] But the main thing to know for our purposes here is that it is very precise. If the wrong base is inserted in the DNA, that mutational error can be passed on to the protein in the form of a mistaken amino acid. Just like a misspelling in a word, a genetic mutation can make a difference. A mistaken amino acid in a protein might cause the protein to fold incorrectly and not work as well, or at all.

Consider an illustration from human language. Let's take the simple sentence, "Methinks it is like a weasel," from Shakespeare's play *Hamlet*. What if it was mistakenly typed out as "Methinks it is like a weakel"? The final word in the sentence is no longer an English word. The sentence no longer makes sense. It's dysfunctional. In a similar way, we know that mutations can often cause serious problems for living organisms. For example, researchers have identified mutations in a protein involved in fruit fly reproduction that lead to a loss of proper function and, ultimately, premature death at either the pupal or larval stage. Several of these fatal results are caused by a single point mutation—the change of a single nucleotide "letter" in the DNA sequence.[25]

So it would seem obvious that mistakes are bad, right? Oddly, proponents of evolution believe that these same mistakes are the raw materials of evolutionary progress. At the most basic level they propose that these mistakes can lead to improvements in the protein.

Now, based on our everyday experience we might conclude that enormous ingenuity and forethought are required to build something like our oxygen transport system. Yet, according to evolutionary theory, it is all due to a series of useful mutational accidents in our DNA.

This is an extraordinary claim, but rather than dismiss it out of hand, we should see if there is any evidence that would count decisively in its favor. As it turns out, any such evidence is lacking, despite an extremely well-funded and assiduous effort by numerous scientists around the globe over several decades. There remains no detailed road map of how genetic mutations could lead to significant improvement and innovation. There is no documentation of these types of improvements in a step-by-step manner.[26] In the absence of such evidence a healthy skepticism is well warranted.

Supporters of modern evolutionary theory point to a few examples of proteins which, when damaged, provide an advantage under unique conditions. The few such examples that have been discovered, however, tend to be double-edged swords. For instance, a mistake in a protein might allow an organism to resist a particular disease, but that same damaged protein will generally not work as well in its normal function.[27]

One researcher, Douglas Axe, who worked in a lab at Cambridge, in England, calculated that the odds of producing a specific protein of 153 amino acids by chance are, as noted above, 1 in $10^{77}$ (that is 1 out of 1 followed by 77 zeros). To get any functional sequence of that length, the odds aren't much better, about 1 in $10^{74}$. For every one functional sequence of that length, there are an enormously large number of gibberish sequences,[28] an enormous ocean of gibberish to swim around in hoping to randomly bump into the ultra-rare functional sequence that codes for a functional protein fold.

It's so enormous we need an illustration to even begin to get our minds around it. Our galaxy has roughly a hundred billion stars in it, and about $10^{67}$ atoms, so your odds are far better of picking the single winning atom, blindfolded and at random, from all the atoms in the Milky Way, than of nature stumbling by chance upon a single new functional protein fold from scratch.[29]

Also, in order to produce significant biological change, evolution has to stumble upon numerous functional protein sequences. And, in many cases, multiple proteins would have been required to work in coordination with each other to produce any biological function that could be subject to natural selection, such as the many protein complexes and molecular machines found throughout biology. Thus, even if Axe's experiment-based estimate is off by many orders of magnitude, the essential challenge to the Darwinian story remains: the odds of repeatedly stumbling upon functional changes and new forms amidst a sea of nonfunctional possibilities appear to be vanishingly small.

## Inertia

With such evidence against modern evolutionary theory, why do its proponents still hold to it? Answering such a question is, of course, largely conjectural, and motivations are unlikely to be exactly the same among even any two scientists, much less among all proponents of evolutionary theory. But speaking broadly, science is a thoroughly human enterprise, and human foibles get in the mix. Historians and philosophers of science have thoroughly documented that scientists, even highly successful ones, have tended to cling to their long-favored theories even in the face of mounting contrary evidence. As it has been said, change in science tends to come one funeral at a time.[30] This is especially the case in origins science where one is often dealing with inferences about unrepeatable past events rather than with a straightforward experiment that can be replicated in the lab. And it's even more the case when the theory in question, evolution, is regarded by some of its proponents as crucial for underwriting their worldview, namely the idea that there is no creator and that reality is ultimately nothing more than matter and energy.

So how do evolutionary theorists justify their support for the theory in the face of so much uncooperative evidence? Those among their number who acknowledge the problems raised above say they just need more time, that we should accept the theory of evolution while science searches for the missing evidence—that to do otherwise is "giving up on science."

It's understandable how they might feel this way, but it's not giving up on science; it's abandoning modern Darwinism—or more precisely, the idea that some cluster of purely blind evolutionary mechanisms could produce all the diversity of life we find around us. It's refusing to let a paradigm trump evidence. By all means, origins biology should continue to pursue promising research, including research into what evolutionary mechanisms can and cannot accomplish. But that is no justification to cling to an inference that has proven wholly inadequate to explain the origin of fundamentally new forms and information.

## Uncommon Descent

RECALL THAT modern evolutionary theory holds that life began with a single-celled organism that in turn evolved into other new forms, and so on until we get to the diversity of living forms we see around us. The idea is commonly illustrated by a gradually branching tree of life. But which forms branched from which ancestors? What did cats evolve from? Are bears and cats closely related? Distantly related? What did turtles evolve from? What about whales? You see, even if one assumes common descent of all living things, there remain thousands of questions about where particular species belong on the evolutionary tree of life. In trying to fill in these details, scientists use living and fossil forms, and they look for similar features across forms. The features used for comparison may be genetic, biochemical, or morphological.

They also make an assumption, namely that similarities often show evolutionary kinship. One could remain open to the possibility that a given similarity between two separate living forms was based on intentional design, in the same way that, for instance, wheels are found on many different kinds of vehicles, not because one evolved into the other but because wheels are a useful design strategy in a wide variety of contexts. But modern Darwinists rule the design possibility out of court and insist that similarities among different living forms cannot be explained as a common design strategy.

As a point of clarification, evolutionists do not assume that every similarity between species is due to a common ancestor that possessed that feature. Evolutionists make exceptions for what they call convergent evolution. These are cases where it is believed that the evolutionary process invented the common feature in question more than once in the history of life; for instance, the similar body forms of fish and dolphins, which are claimed to have evolved separately.

But generally, similarities are assumed to be caused by a shared ancestor that possessed the trait in question. And roughly speaking, the more similar two species are, the more recently their shared ancestor is thought to have lived. So, for instance, the most recent shared ancestor of bobcats and African lions is thought to have lived much more recently than the most recent shared ancestor of lions and bears.

The data sets for constructing an evolutionary tree are by their nature incomplete and subject to interpretation. That by itself isn't a fatal weakness in the argument for common descent by unguided evolution. There is, however, something that may be: radically different and contradictory evolutionary trees can be generated depending on which data sets are used.

When evolutionists construct evolutionary trees using the forms (the morphologies) of plants, animals, and microbes they end up with something quite different from what results when they construct an evolutionary tree based on comparisons of DNA sequences. Even the various trees based on DNA sequences often conflict with each other. Our genomes are huge and complicated, and to make the task of comparing things manageable, evolutionary biologists focus on one particular area of the genome and compare that part across different species. So one genetic evolutionary tree will be based on one part of the genome. A second on another. A third one on another. And so on. Each produces its own evolutionary tree, and these trees can contradict one another, often dramatically so.

Although many proponents of evolution still hold out hope of finding a consistent tree of life, that now seems unlikely. The trees are proliferating rather than beginning to converge on a single true tree of life.[31] To take just one dramatic example of the problem, and as bioengineer Matti Leisola and his co-author Jonathan Witt have noted, a 2013 paper published in the prestigious journal *Nature* "highlighted the extent of the problem. The authors compared 1,070 genes in twenty different yeasts and got 1,070 different trees."[32]

What are we to make of this proliferation of conflicting trees? If all life does indeed share a common ancestor, there is one and only one actual evolutionary tree. It's reasonable to see some conflict as scientists try to build ever more accurate means of discovering what the actual branching history of evolution was, but if evolutionary theory is true, we shouldn't expect to find this ever-expanding forest of conflicting trees. Such a trend makes better sense if the many diverse forms of life are not in fact related by common ancestry, but instead by common design. They could simply share some common design features because they made good design sense to their maker, much as cars, planes, and bicycles share many common design features even though none evolved blindly from the other.

Notice, by the way, that one doesn't have to take an extreme position here. It could be that intelligent design explains many of the recurring design themes we find in biology across different species, and that common ancestry explains commonalities among closely related species— the cat family, for example. That is, maybe all of the different cat varieties we find on Earth did in fact all descend from a single cat ancestor. But the evolution of all life from a single common ancestor through mindless natural processes? The comparative data cast a reasonable doubt on that idea.[33]

## Final Thoughts

WHEN WE think about all the complicated structures found in living organisms, it boggles the mind. Many of these structures possess numerous essential parts, all of which must be present and in place for the system to function. This is known as irreducible complexity. In addition, many of these systems must work together with other irreducibly complex systems for the organism to survive—an irreducibly complex system of irreducibly complex systems. Not only are these coordinated systems consistent with intelligence, but their incredible sophistication also displays a level of genius that far exceeds that of the most brilliant team of human engineers alive. There is good evidence to conclude that unguided processes like those of Darwinian evolution could not produce even a small portion of one of these systems, let alone all of them in their complete and coordinated form.

Some have attempted to refute irreducible complexity by calling upon hypothetical processes like co-option or neutral evolution, or by pointing to some similarities between biological systems, but none of these responses identify a cause adequate to produce fundamentally new biological forms and information.

Others simply dismiss the problems. They say either that they have other evidence for evolution, or that evolutionists will eventually figure things out. Certainly science progresses slowly at times and patience is called for in the stately march of progress, but science often progresses by accumulating evidence against a reigning theory until the only reasonable strategy is to look for an alternative explanation that better fits the evidence.

Supporters of the modern form of Darwin's theory of evolution, neo-Darwinism, hold that natural selection acting on accidental genetic mutations can work creative marvels. But random mutations are, by and large, either neutral (having no observable effect) or damaging. The only caveat is that some damaging mutations do create niche advantages. Fascinating stuff, but it isn't evidence for blind evolution building novel forms and information. It's further evidence that the mutation/selection mechanism is, on its best days, capable only of very modest, degradative variations, like ripping the top off a sedan to create a makeshift convertible.

Those evolutionists who recognize the limitations of the mutation/selection mechanism may still hold to evolution because of the evidence they see for common descent. But common descent is weaker than many suppose. The possession of common features in disparate species is only a strong argument for common descent if common design is ruled out of court from the beginning, prior to a consideration of the evidence. And the evolutionary tree of life that Darwin proposed remains curiously elusive, with various proposed trees jostling for the throne and many contradicting the others.

We are told that mainstream scientists overwhelmingly embrace modern evolutionary theory. But the truth of a scientific theory is not determined by majority vote. And indeed, the history of science is littered with now-discarded theories that were for a time held by the vast majority of scientists in the given field. Also, there are more than a few scientific dissenters from the theory, some of whom work, or have worked, at well-regarded and even prestigious scientific institutions, and who themselves have impressive records of scientific accomplishment.[34]

Given the growing challenges to modern evolutionary theory, perhaps it is time to put intelligent design back on the table of possible explanations and simply follow the evidence where it leads. It may well provide a new way forward toward a better understanding of the remarkable biological systems at the heart of life.

## Review: Your Turn

1. What are some different meanings of the term *evolution?*

2. What evidence, if true, did Darwin acknowledge would seriously undermine his theory?

3. What is irreducible complexity? What is an example from the non-living world? What are some possible examples from biological organisms?

4. Even if we have a complete cardiovascular system, a complete respiratory system, and an interface between the two, what else is needed to ensure that sufficient oxygen is supplied to the body?

5. How have supporters of modern Darwinism tried to respond to evidence-based criticisms of the theory?

6. What are some of the evidential challenges to the idea of common descent?

# Fuel Your Curiosity!

Recommended Resources for Further Exploration:

**VIDEOS**

The Complexity of Life

Bugs with Gears

Untangler of Knots: The Amazing Topoisomerase Molecular Machine

**PODCASTS**

Robert Waltzer on Evolutionary Theory's Room for Humility

The Incredible Design of Muscles
*Jonathan McLatchie*

**ARTICLES**

Charles Darwin: A Short Biography
*Benjamin Wiker*

The Meanings of Evolution
*Stephen C. Meyer and Michael Newton Keas*

**WEBSITES**

*evolutionnews.org*
Original reporting and analysis about evolution, neuroscience, bioethics, intelligent design and other science-related issues, including breaking news about scientific research.

*discovery.org/id/*
The institutional hub for scientists, educators, and inquiring minds who think that nature supplies compelling evidence of intelligent design.

*intelligentdesign.org*
Documents the mounting scientific evidence for nature's intelligent design. Through this site, you can explore the evidence for intelligent design for yourself.

# ENDNOTES

1. For Charles Darwin, there was no goal or purpose in this process; all variations were accidental. Wallace in the end differed in asserting a measure of purposiveness in evolution, as did Asa Gray and others since, but most evolutionary thought has followed Darwin.

2. Darwin himself was unsure whether there was just one original form of life, or a small number of forms (e.g., one ancestor for all plants, one for all animals, one for all fungi), but (at least until very recently) the common view among evolutionary theorists was that there was just one original form.

3. Darwin did not at first call the process of divergence from a common ancestor "evolution" but "descent with modification."

4. For a discussion of some of the difficulties with traditional examples of natural selection, including the peppered moths and Darwin's finches, see Jonathan Wells, *Icons of Evolution: Science or Myth?* (Washington, DC: Regnery Publishing, 2000). See also the companion website at https://iconsofevolution.com/icons-of-evolution/.

5. Yuri Philiptschenko, *Variabilität und Variation* (Berlin: Gebrüder Borntraeger Verlagsbuchhandlung, 1927).

6. Theodosius Dobzhansky, *Genetics and the Origin of Species* [1937] (New York: Columbia University Press, 1982), 12.

7. Winston Ewert, "The Dependency Graph of Life," *BIO-Complexity* 2018, no. 3 (July 17, 2018): 1–27, https://bio-complexity.org/ojs/index.php/main/article/viewFile/BIO-C.2018.3/BIO-C.2018.3.

8. Karl W. Giberson and Francis S. Collins, *The Language of Science and Faith: Straight Answers to Genuine Questions* (Downers Grove, IL: InterVarsity Press, 2011), 43.

9. S. J. Padayatty and M. Levine, "Vitamin C: The Known and the Unknown and Goldilocks," *Oral Diseases* 22, no. 6 (September 2016): 483, https://doi.org/10.1111/odi.12446.

10. See, for example, William A. Dembski, *The Design Revolution: Answering the Toughest Questions about Intelligent Design* (Downers Grove, IL: Intervarsity Press, 2004), 317; see also Jonathan Wells, *The Myth of Junk DNA* (Seattle: Discovery Institute, 2011).

11. Ann K. Gauger, Ola Hössjer, and Colin R. Reeves, "Evidence for Human Uniqueness," in *Theistic Evolution: A Scientific, Philosophical, and Theological Critique*, eds. J.P. Moreland et al. (Wheaton, IL: Crossway, 2017), 497. [An internal citation was removed from the quotation.]

12. Charles Darwin, *The Origin of Species*, 6th ed. (London: John Murray, 1872), chap. 6, http://darwin-online.org.uk/content/frameset?itemID=F391&viewtype=image&page seq=1.

13. Michael Behe, *Darwin's Black Box: The Biochemical Challenge to Evolution*, 10th anniversary ed. (New York: Free Press, 2006), 39.

14. Behe, *Darwin's Black Box*, 39 (for the quoted phrase), 42–45 (for the discussion of the mousetrap's parts and their interactions).

15. Behe, *Darwin's Black Box*, 51–139.

16. Scott A. Minnich and Stephen C. Meyer, "Genetic Analysis of Coordinate Flagellar and Type III Regulatory Circuits in Pathogenic Bacteria," in *Design and Nature II*, eds. M. W. Collins and C. A. Brebbia (Southampton, UK: WIT Press, 2004), 302. See also the short online video, "Type Three Secretory System," https://revolutionarybehe.com/category/bacterial-flagellum/.

17. Some of these ideas are taken from an excellent interview with Dr. Howard Glicksman, "A Doctor Examines How the Body Meets Its Need for Oxygen," September 27, 2017, in *ID the Future*, podcast, MP3 audio, 17:19, https://www.discovery.org/multimedia/audio/2017/09/a-doctor-examines-how-the-body-meets-its-need-for-oxygen/.

18. Stuart Fox, *Human Physiology*, 15th ed. (New York: McGraw-Hill Education, 2018), 408.

19. According to one analysis, the size of each hemoglobin molecule is approximately 5 nm in diameter. See Harold P. Erickson, "Size and Shape of Protein Molecules at the Nanometer Level Determined by Sedimentation, Gel Filtration, and Electron Microscopy," *Biological Proceedings Online* 11, no. 1, art. 32 (May 15, 2009): 35, https://biologicalproceduresonline.biomedcentral.com/track/pdf/10.1007/s12575-009-9008-x.

20. Douglas Axe, "Estimating the Prevalence of Protein Sequences Adopting Functional Enzyme Folds," *Journal of Molecular Biology* 341 (2004): 1295–1315.

21. Bruce Alberts et al., *Molecular Biology of the Cell*, 6th ed. (New York: Garland Science, 2015), 16.

22. For more information on this evolutionary constraint, see Ann K. Gauger, Stephanie Ebnet, Pamela F. Fahey, and Ralph Seelke, "Reductive Evolution Can Prevent Populations from Taking Simple Adaptive Paths to High Fitness," *BIO-Complexity* 2010, no. 2 (January 2010): 1–9, http://dx.doi.org/10.5048/BIO-C.2010.2.

23. To understand the reason behind the guarded language here, see Mark Gerstein et al., "What Is a Gene, Post-ENCODE?" *Genome Research* 17 (2007): 669-681, https://genome.cshlp.org/content/genome/17/6/669.full.html, as well as Jonathan Wells, *Zombie Science: More Icons of Evolution* (Seattle: Discovery Institute Press, 2017), chap. 4.

24. For a simple explanation and an early animation of protein synthesis, watch *Unlocking the Mystery of Life*, directed by Lad Allen (La Mirada, CA: Illustra Media, 2003), DVD, chapter 10. This portion of the documentary is also available for viewing: Illustra Media, "Unlocking the Mystery of Life (Chapter 10 of 12)," YouTube, video, 4:27, December 9, 2008, https://www.youtube.com/watch?v=gdBJt6sdDfI.

25. For a detailed discussion of several mutations in the flapwing protein (flw) and the impact on fruit fly reproduction, see Shinya Yamamoto et al., "Protein Phosphatase 1ß Limits Ring Canal Constriction during *Drosophila* Germline Cyst Formation," *PLOS ONE* 8, no. 7 (July 25, 2013): e70502, https://doi.org/10.1371/journal.pone.0070502.

26. Michael Behe's follow-up to the bestselling *Darwin's Black Box* examined the experimental evidence and field observations to determine what Darwin's mutation-plus-selection mechanism could actually accomplish. He looked at microbes, since they have huge populations and rapid generational turnover, allowing evolutionary processes to try many millions of mutations over just a few years. From the results he extrapolated mathematically to even longer waiting times and still larger populations. From this work he showed that there are severe limits to the ability of the Darwinian mechanism to effect biological change. It can tinker but not innovate. It can break but not build anything fundamentally new. See Michael J. Behe, *The Edge of Evolution: The Search for the Limits of Darwinism* (New York: Free Press, 2007).

27. Behe has shown that the Darwinian mechanism is most effective as a destructive force, rather than as a creative one. Behe set forth a principle he calls "the first rule of adaptive evolution," which, in essence, states that mutations that yield a net fitness gain are much more likely to be mutations that break or blunt a pre-existing function than ones that produce a new function. See Michael J. Behe, "Experimental Evolution, Loss-of-function Mutations, and 'The First Rule of Adaptive Evolution,'" *The Quarterly Review of Biology* 85, no. 4 (December 2010): 419–45, https://www.lehigh.edu/~inbios/Faculty/Behe/PDF/QRB_paper.pdf. See also his recent book, *Darwin Devolves: The New Science about DNA That Challenges Evolution* (New York: HarperCollins, 2019).

28. Douglas Axe, "Estimating the Prevalence of Protein Sequences Adopting Functional Enzyme Folds." See also Douglas D. Axe, *Undeniable: How Biology Confirms Our Intuition That Life Is Designed* (New York: HarperOne, 2016), 57.

29. Some researchers think the unlikelihood of accidentally forming a protein is even higher than Axe's calculations focused on amino acid sequences. Other factors also required for functional proteins include the free energy states of the amino acid combinations and the stability of the folded protein chain. Recent research hints at additional improbabilities similar to those calculated by Axe. See Brian Miller, "Thermodynamic Challenges to the Origin of Life," in Charles B. Thaxton et al., *The Mystery of Life's Origin: The Continuing Controversy* (Seattle: Discovery Institute Press, 2020), 359–74.

30. The original statement was: "A new scientific truth does not triumph by convincing its opponents and making them see the light, but rather because its opponents eventually die and a new generation grows up that is familiar with it." From Max Planck, *Scientific Autobiography and Other Papers*, trans. Frank Gaynor (London: Williams & Norgate, 1950), 33–34. Others have taken Planck's statement and modified it for brevity, as seen in Pierre Azoulay, Christian Fons-Rosen, and Joshua S. Graff Zivin, "Does Science Advance One Funeral at a Time?," *American Economic Review* 109, no. 8 (2019): 2889–2920. Similar ideas are presented in Thomas Kuhn, *The Structure of Scientific Revolutions*, 4th ed. (Chicago: University of Chicago Press, 2012).

31. See Stephen C. Meyer, *Darwin's Doubt: The Explosive Origin of Animal Life and the Case for Intelligent Design* (New York: HarperOne, 2013), chap. 6.

32. Matti Leisola and Jonathan Witt, *Heretic: One Scientist's Journey from Darwin to Design* (Seattle: Discovery Institute Press, 2018), 84. The paper referred to is Leonidas Salichos and Antonis Rokas, "Inferring Ancient Divergences Requires Genes with Strong Phylogenetic Signals," *Nature* 497 (May 16, 2013): 327–31, https://doi.org/10.1038/nature12130.

33. See Günter Bechly and Stephen C. Meyer, "The Fossil Record and Universal Common Ancestry," in *Theistic Evolution: A Scientific, Philosophical, and Theological Critique*, eds. J. P. Moreland et al. (Wheaton, IL: Crossway, 2017), 331–62

34. "Dissent from Darwin List Tops 1,000—Now the Scientists Weigh In," *Evolution News*, February 14, 2019, https://evolutionnews.org/2019/02/listen-dissent-from-darwin-list-tops-1000-scientists-weigh-in/.

# IMAGE CREDITS

**Figure 1.** Tree of Life. "Genealogical Tree of Humanity." Illustration by Ernst Haeckel, ca. 1877. Modified by Fuelbottle, 2007, Wikimedia Commons. Public domain.

**Figure 2.** Galapagos finches. Illustration by John Gould in Charles Darwin, *Journal of Researches into the Natural History and Geology of the Countries Visited during the Voyage of H. M. S. Beagle round the World* (London: John Murray, 1845), 379. Modified by Shyamal, Wikimedia Commons. Public domain.

**Figure 3.** Common mousetrap. Image by Eric H. Anderson.

**Figure 4.** Electron micrograph of a bacterium. Transmission electron micrograph by Graham Bradley, 2005, Wikimedia Commons. Public domain.

**Figure 5.** Bacterial flagellum. Illustration by Joseph Condeelis/Light Productions. Adapted by Brian Gage.

**Figure 6.** Alveolar sacs and pulmonary capillaries. Image by LadyofHats and Salman666, 2007, Wikimedia Commons. Public domain.

# WHAT IS THE DISCOVERY SOCIETY?

The Discovery Society is a group of individuals who come together to support the work–and disseminate the message–of Discovery Institute's Center for Science and Culture. New members receive materials that help educate themselves and spread the word about our work to those in their circle of influence. Depending upon their giving level, members receive one to three Discovery Institute Press newly released books per year, along with invitations to regional donor events and discounted rates on our annual Insiders Briefing events.

If you appreciate this booklet and aren't already a member, we hope you will consider joining our network of supporters today!

Your donation to Discovery Institute's Center for Science and Culture will allow us to expand our cutting-edge scientific research and scholarship; train young people through our education and outreach; and reach the masses through media and communications.

*discovery.org/id/donate*

MORE INFORMATION ON THE DISCOVERY SOCIETY CAN BE FOUND AT
*discovery.org/id/donate/#member-levels.*

# Evolution and Intelligent Design in a Nutshell

Are life and the universe a mindless accident—the blind outworking of laws governing cosmic, chemical, and biological evolution? That's the official story many of us were taught somewhere along the way. But what does the science actually say? Drawing on recent discoveries in astronomy, cosmology, chemistry, biology, and paleontology, *Evolution and Intelligent Design in a Nutshell* shows how the latest scientific evidence suggests a very different story.

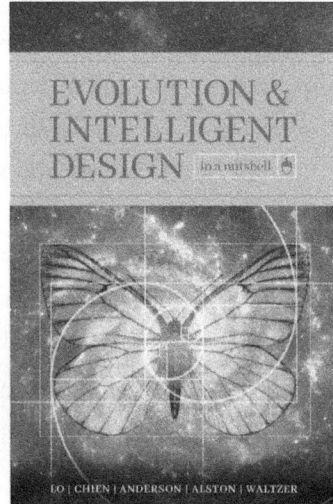

*"accessible, informative... powerful ... an excellent resource."*

J. Warner Wallace

## PURCHASE THE FULL BOOK HERE:

*DiscoveryInstitutePress.com/EvolutionandID*

# MORE IN THIS SERIES:

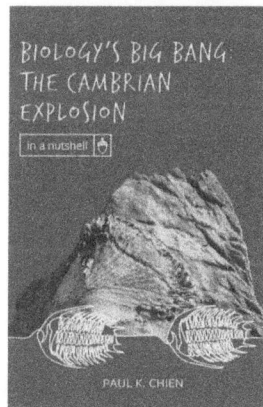

THE BIG BANG &
THE FINE-TUNED
UNIVERSE
in a nutshell
ROBERT A. ALSTON

THE ORIGIN OF LIFE
& THE INFORMATION
PROBLEM
in a nutshell
ERIC H. ANDERSON

FACTORIES THAT
BUILD FACTORIES
in a nutshell
ERIC H. ANDERSON

BIOLOGY'S BIG BANG
THE CAMBRIAN
EXPLOSION
in a nutshell
PAUL K. CHIEN

This series of booklets was created to help Discovery Society members educate themselves about the basic arguments for intelligent design and the critiques of Darwinian evolution. Each booklet presents the content of one chapter of *Evolution and Intelligent Design in a Nutshell*. To help you delve deeper into each subject, we have included a list of recommended resources from our vast library of videos, podcasts, articles, and websites. Members of the Discovery Society can download digital versions of these books through the Discovery Society Community on the DiscoveryU.org platform or purchase physical copies at a discounted rate through Amazon.com.

www.ingramcontent.com/pod-product-compliance
Lightning Source LLC
Chambersburg PA
CBHW022054190326
41520CB00008B/785